Unveiling Disease X:

Understanding and Getting Ready for an Emerging Global Infectious Threat

Table of Contents

Introduction

Health conditions generated by microorganisms like bacteria, viruses, fungus, or parasites are known as infectious diseases. Numerous species live inside of our bodies. In most cases, they are beneficial and sometimes safe. But in specific circumstances, some bacteria have the capacity to trigger global disease.

It is possible for some infectious diseases to spread from one individual to another. Others are spread by animals or insects. And you could contract others if you consume infected food or water or come into contact with organisms from the environment.

Fever and weariness are frequent symptoms and indications of infection, though they might differ depending on the pathogen that is triggering it. While severe life-threatening illnesses may require being hospitalized, infections that are mild may be treated with rest and home treatments.

Vaccination assist in avoiding a lot of infectious diseases, including measles and chickenpox, coughing, fever, diarrhea, and muscle aches.

Viruses That Have the Potential to Begin a Pandemic

The following virus are likely to start a pandemic;

Lassa Virus

The infection known as Lassa fever is an invasive hemorrhagic disease that breaks down blood vessels and destroys internal organs. It is caused by Lassa virus.

One in five Lassa virus infections results in a life-threatening illness that affects the liver, kidneys, or spleen. The virus is frequently spread through infected household items through rat's pee or feces. Health professionals who come into proximity to a patient's blood or tissue from an organ may also contract an infection. After recovering from a lassa fever illness, loss of hearing may persist. Lassa fever frequently produces extended epidemics in numerous West African countries, where 1–16% of people are infected, unlike Ebola and Marburg, which occasionally generate lethal outbursts with a subsequent decline.

Ebola Virus

Though it seems like the most severe Ebola outbreak was in the nation of Democratic Republic of Congo (DRC. Considering the swift spread of the disease across numerous nations and the concern about a further devasting spread, this case, together with the West African Ebola epidemic in 2014, spurred an enormous mobilization of resources and money for controlling the pandemic.

Since Ebola can be transmitted by direct contact with infectious fluids in the body, particularly blood, feces, and vomit, it requires intimate interaction with people to spread. Healthcare providers and the relatives of infected individuals are at particularly high risk as a result.

Marburg Virus

Marburg, which is a member of the same virus group as Ebola, has symptoms that are comparable to those of Ebola and is spread similarly through contact with the blood, body fluids, or tissues of those who are sick.

Similar to how Ebola spread, burial customs that call for close touch with the person who passed away increase the risk of disease transmission. Marburg is extremely dangerous, killing up to 87% of those who contract it.

Corona Virus

SARS-CoV-2, also known as COVID-19 or "coronavirus infection 2019," is a transmissible illness that affects humans, it quickly spread over the world and became a pandemic. Whenever a person with the infection sneezes, coughs, or otherwise talks, or inhales air, COVID-19 mainly spreads by airborne droplets.

However, the virus can also be shared by contacting objects that have been exposed to the virus.

Severe Acute Respiratory Syndrome (SARS)

In comparison to COVID-19, SARS, which previously created a global pandemic in 2003, killed 775 people in 37 different countries. However, there is no assurance that the SARS virus won't resurface and do even more harm.

Similar to COVID-19, the virus is believed to spread through respiratory droplets released when sneezing or coughing. However, unlike COVID-19, those who have SARS appear to constantly have severe symptoms, making them simpler to spot.

Nipah Virus

Nipah, a relative of the measles virus, has been responsible for numerous cases in South-East Asia since its discovery in 1998. It is capable of killing up to 77% of those it affects. Severe swelling in the brain brought on by the infection can result in pain in the head, nausea, dizziness, stiff necks, and even coma-like symptoms. Nipah is a frequent disease among bats, notably fruit-eating bats in South-East Asia, and like some of the illnesses on the above list, it first appeared in animals. In close proximity with diseased pigs and foods that have been tainted with infested bats' urine or saliva are the main ways in which it transmits. Transmission may occur via respiratory discharges like sneezing or coughing, though there is a risk that the infectious agent can evolve and become far more contagious.

The Rift Valley

Rift Valley fever is an illness transmitted by mosquitoes that primarily affects livestock, however, it can also infect humans whenever people come in touch with infected animals' fluids from their bodies, such as their blood or milk. Additionally, the bites of mosquitoes might spread the disease to them. There hasn't been any evidence of transmission from one person to another. When it comes to humans, a simple infection can result in a high temperature and pain in the muscles, while more serious infections might result in lack of vision, brain damage, or bleeding that is uncontrolled. Rift Valley disease epidemics have been documented for decades in a number of African nations, and since 2000, they have also spread to the Middle East.

Zika Virus

Most illnesses brought on by the Zika virus are modest, involving a high temperature, skin eruption, and muscle soreness. However, in 2015–2016, Zika triggered a frightening occurrence of birth malformations known as fetal Zika syndrome: Infants delivered to pregnant Zika-infected women are at danger for microcephaly in as well as a greater likelihood of miscarriage.

Researchers have lately noticed that infants who were formerly born without any obvious symptoms may subsequently show issues including eyesight loss. The bites of Aedes mosquitoes, which also transmit the fever and chikungunya viruses, are the source of the zika virus.

Crimean-Congo hemorrhagic fever virus (CCHFV)

Hemorrhagic fever is typically transmitted by being bitten of an infected tick, this disease mostly impacts animals, mainly livestock. However it can also infect humans if they come into touch with freshly butchered animals that have been infected. There have been occasional cases of transmission from person to person through contact with an infected person's blood or urine.

The infection first causes symptoms resembling the flu and may occasionally induce light sensitivity or neck stiffness that may be mistaken for measles, but subsequently patients may experience acute, excessive bleeding. Due to the presence of the particular kind of tick that transmits it.

Monkeypox Virus

17

Smallpox-like symptoms, such as a broad pustular rash, a high body temperature and tiredness, are caused by monkeypox virus. However, it can also be passed from one person to another by coming into contact with wounds, bodily fluids, respiratory droplets, contaminated clothing or bedding, and exposure to wild animals like mice and monkeys.

The newly developed third-generation vaccine has now been licensed for avoiding the spread of monkeypox, and the vaccine that was utilized for the elimination of smallpox can now provide protection against monkeypox.

Monkeypox has traveled from Africa's central and western regions to North America and other European nations including the UK due to travel abroad and pet trafficking, but all of the subsequent cases have thus far been controlled.

What's Happening that's Causing New Infectious Risks to Emerge?

The formation of novel infectious illnesses or the resurgence of old infectious illnesses is influenced by a variety of circumstances. While some are the product of human activity and activities, many are the outcome of natural processes, such as the emergence of viruses over time.

Think about how, particularly during the past century, the relationship between humanity and the environment has altered. Population expansion, movement of people from the countryside to cities, traveling abroad, poverty, conflicts, and detrimental changes to the environment brought on by economic growth and agricultural practices are among the variables that have led to these shifts.

There must be no fewer than two circumstances for a newly identified illness to be officially established. A vulnerable population must be exposed to the contagious agent, and the pathogen has to be capable to transmit easily from person to person and cause illness. Additionally, the illness must be able to spread across the general population in order to let a growing number of individuals come into contact with it.

Whenever infectious organisms from animals spread to people, it is known as a zoonose, and this causes a number of new diseases.

The likelihood that individuals are going to come into close proximity with different kinds of animals that are possible hosts of an infectious disease. It rises as the population of humans grows and spreads into territories that are unfamiliar. It is clear that this combination of factors poses a significant risk whenever that element comes together with rises in population size and movement.

Growing concern exists regarding the role that climate change may have in the propagation of infectious diseases.

Diseases can travel to new regions when the climate on Earth increases and habitats change. For instance, rising temperatures enable mosquitoes and the illnesses that they spread to spread into areas where they were not formerly present.

The inherited resistance of infections to antimicrobial drugs like antibiotics, is an aspect that is particularly significant in the resurgence of diseases. Medicines used for the treatment of diseases brought on by infections can become less effective as time progresses due to changes in viruses, bacteria, and various other organisms. Medications that were once useful for treating disease have therefore become becoming less effective.

A drop in vaccination rates, wherein an increasing number of people decide not to get vaccinated regardless of whether a safe and effective vaccine is available, can also lead to a disease's reemergence. This appears to be an issue with the measles vaccine especially.

Measles, a severe and highly transmittable illness that was eradicated from the western part of the globe in 2016 and the United States in 2000, is now back in some areas as a result of a rise in the percentage of people choosing nonmedical vaccine exemptions because of their private and religious convictions. This phenomenon was fueled by an anti-vaccination campaign that was largely predicated on a false and debunked study that suggested a connection between the measles vaccine and Autism Spectrum Disorder.

Guidelines for Leading a Healthy Lifestyle to Improve Immune System Function

There are a number of pills and products on the market that promise to increase immunity. However, maintaining an effective immune system requires more effort than simply ingesting a combination of mineral and vitamin supplements in the form of a tablet or powder.

The immune system in your body maintains a very precise equilibrium while functioning. It needs to be powerful and intelligent enough to fend off a wide range of illnesses and infections, nevertheless not so powerful that it over responds excessively.

It is really highly regulated to accomplish this by a variety of inputs as well as in response to what's going within your body. A number of measures you can try to assist your body's defense system with what it needs to perform efficiently, though, whether you're battling a cold, or COVID-19, or flu.

Here are methods to develop and sustain a powerful, robust immune system, all supported by science:

- **Keep abreast with the suggested vaccinations**

Having an effective immune system includes using vaccines, which provide us the best chance to defend ourselves against dangerous diseases.

The immune system of a person is intelligent, but vaccinations help it become even more so by teaching it how to identify and combat particular diseases. Your body's defenses acquires knowledge much more safely through vaccination than through direct contact with these potentially dangerous pathogens.

- **Regular exercise**

Exercise is crucial for maintaining a healthy state and an effective immune system, in addition to enabling you grow muscles and reduce stress.

Exercising at moderate level causes antibody-producing cells to be released from the bone and enter into the bloodstream. It additionally helps in the transfer of immune cells that are already inside the bloodstream toward the tissues. This enhances immune monitoring. Immunological monitoring can be thought of as a mechanism that keeps an eye on your defenses, with exercise enhancing immune cells' capacity to recognize and respond to illness. This implies that maintaining activity levels and engaging in regular exercise should be a priority.

- **Get lots of rest**
  Even though it might not seem like it does, the human body is actively working while you sleep on a number

of important functions. Sleeping is crucial for the body's immune homeostasis and immunological function.

Knowing how many hours of sleep you should receive each night and what to do if it isn't happening is crucial for giving the body's defenses the greatest possible chance to fend off illness and sickness.

- **Keep a balanced diet**
  A nutritious diet is essential for an efficient immune system, as it's associated with most other aspects of your body. To do this, make sure you consume an abundance of vegetables, legumes, whole grains, zinc, fruits, lean meats, and healthy fats.

Your defense system's homeostasis is maintained whenever your body gets enough the micronutrients included in the aforementioned foods.

Vitamin B6, which can be found in chicken, fish such as salmon and tuna, green vegetables, bananas and potatoes is one of these essential. Citrus fruits which includes berries, oranges, tomatoes, broccoli, among others all contain vitamin C. The greatest strategy to maintain your immune system is to consume a well-balanced diet because researchers think that the human body absorbs vitamins more effectively from food sources compared with supplements.

- **Hydrate yourself thoroughly**
  Your body needs water for a variety of vital functions, including immune system support.

  Water is essential because our blood and lymph nodes, which contain cells that fight infection, require water for them to circulate throughout our bodies.

  You regularly lose water through the air you inhale, urine, and elimination of stool, regardless of whether you're not working out or perspiring. Make sure you're replenishing any water you lose through sweat that you can utilize to strengthen your body's defenses, which begins with understanding how much water you ought to be drinking each day.

The Importance of Consistent Handwashing, Respiratory Manners, and Good Hygiene

In order to protect the health of everyone, frequent washing our hands, proper breathing hygiene, and proper sanitation are of utmost importance. The washing of hands is an essential part of illness prevention yet is frequently overlooked.

By thoroughly washing hands using soap and water, dangerous microorganisms are loosened and removed, reducing the spread of infections, from typical colds to more serious illnesses such as COVID-19.

When someone coughs or sneezes, they should cover their nose and mouth to stop any droplets that are infectious from entering the air, which helps to stop the propagation of respiratory infections. Such straightforward measures can significantly reduce the spread of infections and protect vulnerable populations, especially youngsters and seniors.

Maintaining good sanitary standards goes beyond just protecting one's health, bathing frequently and keeping your home clean both contribute to the creation of a climate that is less favorable for the spread of bacteria and viruses. The result of this group effort is a decrease in the total expense of diseases to health care systems and the promotion of a more healthy population.

To also safeguard the population at large, it is essential to get personal first aid items such as hand sanitizers when you come in contact with people.

Being Aware Of the Value of Immunizations and How They Help Fight Against Infectious Diseases

One of the most effective methods to shield yourself, your family, as well as future generations of people against infectious diseases is through vaccination. To put it another way, by getting vaccinated, you can prevent the transmission of diseases both today as well as in the years to come. You are taking care of yourself and your family as well as vulnerable others in your local area through making sure that you and everyone in your household are completely immunized. An illness may propagate more slowly and with fewer people being affected as a result of increased vaccination rates. Immunizations helps to prevent death caused by infectious diseases.

The effects of each immunization are the same. The immunization boosts your immune system's capacity to fight off infections before you even come into close proximity to them. It is similar to having the condition but not showing signs or symptoms.

After receiving a vaccination, if you happen to come into touch with an infectious agent, your body will strive to prevent you from contracting the illness, or you may only experience a minor case. Vaccines have undergone extensive testing to show that they are both safe and efficient at preventing diseases that are transmissible.

Some people are resistant to some vaccines. They are possibly prematurely old or too ill, which could explain this. You can help to safeguard these helpless people by having your family's vaccines current. You significantly increase your own and your community's safety by being vaccinated.

Since disease cannot travel readily from one person to another, when enough members of the community are immunized, disease transmission is slowed or halted entirely. The sickness won't spread as long as many individuals receive the vaccine.

Friends, relatives, and others are safeguarded by what is known as community immunity, particularly those who cannot receive immunizations.

Preparing How to Communicate and Work Together With Close Family Members during a Pandemic

During an infectious outbreak, communication is essential. Think about scheduling routine video chats, starting an online chat sharing updates, and exchanging contact information for emergencies. Discuss safety precautions and crucial details to share, as well as alternate arrangements in case someone gets sick.

Maintaining communication with close relatives is crucial during emergencies like an outbreak. An organized strategy for communication and coordination may provide people with a sense of safety and guarantee that everybody is well-informed and supported.

Create a communication center first. Consider a group conversation on a messaging program like WhatsApp or Telegram if you want an interface which functions well for all.

This hub will act as a focal point for exchanging information, updates, and reports on one another's health. Keeping everybody in the loop and removing uncertainty will be made easier with continuous communication.

Think about setting up routine video calls.

Even when two people are physically separated, being able to see their respective faces can help them feel connected. Make these phone calls at specific times so that everybody can schedule their days around them. Phone conversations not only provide emotional assistance, but also a chance to exchange crucial details and condition information.

Provide your family's emergency contact information. Make sure that everybody has access to contact information, locations, and any necessary health-related data. Name the primary contact who may be contacted if anyone requires help or advice.

This is especially useful if a loved one becomes unwell and needs assistance.

Decide on safety regulations after discussion. Ensure that everybody is cognizant of the advised measures and distribute information from reliable sources, such as the official websites of health organizations.

Promote open communication regarding the epidemic, clarifying any worries, and making sure that everybody is aware of how serious the issue is.

Finally, make backup plans. While planning for different outcomes, it is a good idea to hope for the best. Talk about how to handle daily necessities like shopping and prescription, what to do if a member of the family becomes ill, and how to offer support remotely.

Making a thorough plan may assist to secure the health and safety of everyone by allowing people to communicate and coordinate their efforts.

Decide on a main connectivity route to use first. Perhaps it's an interactive chat on a messaging system or an electronic mail. Pick an approach that everybody in the family feel at ease utilizing. This unified channel will be used to disseminate important news, developments, and vital details about the outbreak.

The secret is regular interaction. Decide on a timetable for meetings, video calls, and online meetings. Continuity will reduce anxiety and give a forum for discussing problems and perspectives. Everyone may stay informed while remaining on the same page by receiving frequently updated information on the outbreak's development, safety precautions, and official instructions.

Techniques for Managing Stress and Anxiety during the Global Infectious Threat

What you should understand about dealing with stress amid infectious disease outbreaks, you may experience anxiety and exhibit indications of stress as soon as you hear, read, or witness news coverage about it.

These stress-related symptoms are common and may even be more prominent in persons who have family members who live in the outbreak's afflicted regions. Keep an eye on your own mental and physical health after the spread of an infectious disease.

Understand how to manage your stress as well as when to get assistance.

The reactions that follows are, mental, physical, and cognitive symptoms of anxiety and stress. After learning about an epidemic of an infectious disease, you might start noticing some of them.

Here a possible behaviors you might exhibit;

- An upsurge or drop in your mental and physical activity levels.
- Regular weeping.
- A rise in your smoking, drinking, or substance abuse.
- Difficulty in interacting with others.

- A rise in irritation, with violent eruptions of frustration and continual arguments.
- A refusal to rest or sleep.
- The need for solitude majority of the time
- Continuous worrying about things.
- Having trouble offering or accepting assistance.

Bodily observations include;

- Having discomfort in the stomach.
- Experiencing head pain and other pains.
- Sweating or experiencing shivers.
- Reducing the need to eat or overeating.
- Having tremors or twitches in your muscles prone to becoming startled.

Awareness sensation;
- Making judgments is challenging understand.
- Having problems recalling details.
- Feeling perplexed and having difficulty focusing and thinking clearly.

Emotional sensation;
- Not giving a damn about anything
- Experiencing depression
- Being nervous or frightened
- Feeling enraged
- Having a heroic, joyous, or invulnerable feeling.

You Can Control and Reduce Your Stress Using the Following Strategies

- **Verify The Truth:**
  Look for sites as well as individuals you can trust to provide reliable health information. If you are in danger, ask them for information on the current outbreak along with how you may safeguard yourself from sickness. Your family's physician, a state or neighborhood healthcare, the national government offices

- **Remember To Keep Everything in View**

Create time restrictions on how much you may read or see regarding the outbreak of disease. If you or someone you know have family members or loved ones living in areas where the outbreak of disease has been widespread, you are bound to be informed of any fresh information. But be sure to set out some time to concentrate on the aspects of your existence that are progressing well and that which you have control over.

- **Apply Practical Means Of Rest**
  Schedule yourself between strenuous activities, and relieve yourself after a difficult assignment by doing what you find relaxing: deep breathing, stretching, meditation, washing your

face and hands, or engaging in enjoyable hobbies are activities you can do. Spend your free time relaxing, have a satisfying dinner, read a book, enjoy good music, have a bath, or speak to friends.

- **Maintain Good Health**
Eat healthy foods, take water, engage with some workout. Avoid excessive caffeine as well as alcohol. Don't smoke or use illicit substances. Get enough rest and sleep.

- **Communication with others**
Make contact with those who are possibly under stress due to this outbreak. To help you remember the numerous significant and uplifting

aspects that define your lives, speak about how you're feeling regarding the outbreak, provide reliable medical advice, and engage in discourse unconnected to the outbreak. Spend a few moments in prayer, meditation, or serving others who are in a position to refresh your mind and soul.

Conclusion

Furthermore, there are various procedures and recommendations that ought to be implemented to reduce the transmission of infections amongst persons, within a group, and internationally. Prevention and control of infection is a global priority. Recognizing populations that are vulnerable, including kids, the elderly, and individuals with chronic illnesses, may assist in directing initiatives to safeguard these particular populations.

A community-wide approach to prevention of infection can begin by altering behaviors, such as Habitual sanitation of hands, recommended utilization of facial masks, applying repellents to avoid insects, ensuring regular immunizations are current, and taking part in immunization programs, adhering to medical professionals' recommendations for administering prescription drugs, especially antibiotics.

Bonus Package

This is a free gift from me to you; it is a manual on how to make liquid soap and hand sanitizer locally, which is a crucial hygiene tip for staying safe from infectious diseases at all times.

Why you need to utilize this package

People may not have easy accessibility to stores or purchasing goods online when they are instructed to remain at home or when there is a lockdown during a global infectious threat.

You can maintain can be free from infectious threat if you have the knowledge and skills to make these products for yourself and close acquaintances.

How to Make an Affordable Hand Sanitizer that Will Prevent Infection Spread

Scientific research has shown that when people come into contact with one another, touch objects like windows, doors, tables, chairs, etc., microscopic organisms like viruses, fungi, and bacteria can live on these surfaces and spread to our bodies via our hands. Only a magnifying lens can be used to see microorganisms, which are small creatures.

Every surface that they come into contact with will likely result in their reproduction and spread. To help with the process of multiplication on surfaces, they both have male and female reproductive organs.

If not successfully handled, this causes illnesses and diseases in humans and can result in mortality.

An antibacterial liquid known as hand sanitizer is intended to be applied and on the hands in order to get rid of the microorganisms. When those agents are on our hands, there is a risk that they will spread to any region of the body that we touch.

We humans are advised to use hand sanitizer after coming into contact with other people and objects.

In particular, hand sanitizers are required in establishments including educational institutions, medical facilities, churches, mosques, airports, and others to reduce the likelihood of viral transmission.

Government regulations require the use of hand sanitizers in public areas where people socialize due to the pandemic's spread.

Equipment Required for the Production of Hand Sanitizer

The equipments needed for producing hand sanitizer are listed below, along with an in-depth description of each tool.

1. Mixing dish

The mixture is blended in a mixing dish. It contains the proper amounts of chemicals required for the production of hand sanitizers. This means that you need to purchase bowls big enough to hold the mixture, depending on how much you want to make.

2. Stirring Stick

Products are stirred or mixed with stirring sticks. It is crucial to remember to stir chemicals in single direction (either clockwise or anticlockwise). Avoid stirring in the opposite direction since the substance can splash on your skin.

3. Nose Mask

In order to avoid inhalation of the chemicals utilized in the manufacture of hand sanitizer, a mask for the nose is crucial. Since the human system is unique, Individuals may all be allergic to inhaling certain substances, which could be harmful to our health.

4. Measurement Device

To make hand sanitizers, a certain amount of certain chemical is required. Your measuring device will help you determine the appropriate number and proportions to use.

Note

Chemical amounts should be followed in the proper proportions to prevent incorrect production. Certain individuals have sensitive skin to certain substances. It is recommended to use the accepted measurements.

Multiply each quantity by the number you want to receive if you are interested in doubling or tripling the quantity that you would like to produce.

Chemicals Required for Making Hand Sanitizer and their Purposes

The following substances are required to make hand sanitizer.

1. Ethanol or Isopropyl alcohol

2. Aloe vera

3. Mint pepper oil.

4. Perfume.

5. Glycerin

Ethanol or Isopropyl Alcohol

Alcohol's anti-bacterial properties render it necessary and useful. It aids in the destruction of tiny organisms that might live on the body.

Aloe Vera Gel

Aloe vera gel is utilized to make hand sanitizer because it helps alcohol kill microorganisms. To our eyes, it also gives the hand sanitizer a thicker appearance.

3. Peppermint oil

Several individuals' skin is hypersensitive to the peppery sensation, hence peppermint oil is an optional ingredient in hand sanitizer production.

4. Perfume oil

The hand sanitizer has a lovely scent thanks to the fragrance oil. There are numerous accessible fragrances.

5. Glycerin

Glycerin aids in hand moisturization. It draws a small amount of water from the hand's inside, which settles on the outermost layer of the skin.

Chemicals Required to Produce Hand Sanitizer in Quantities

- One liter of Ethanol or Isopropyl alcohol.
- 1/2 kilogram of Aloe vera
- 10 ml of peppermint oil
- 20ml of fragrance
- 10 ml of glycerin

Production of Hand Sanitizer through Mixing

1. Prepare a fresh mixing bowel.
2. Carefully stir the alcohol into the Aloe vera gel in the amount mentioned above until the texture is as desired.
3. Stir in one direction after adding a small amount of peppermint oil.
4. Add 10 ml of glycerin and carefully stir in a single direction.

5. Add the scent and mix.

6. Package the product.

Safety Measures during Hand Sanitizer Manufacturing

Take note of the following safety measures when making hand sanitizers.

1. **Keep chemicals away from youngsters**

Such substances can be mistakenly viewed as amusing by children. So it was necessary to keep chemicals away from them.

2.   When producing, wear a nose mask to avoid inhaling chemicals that your body can be sensitive to.

3. Make sure to stir mixtures in a single direction when doing so to prevent chemical splashes on the surface of the body.

How to Make an Affordable Liquid Soap for Handwashing

Liquid soap is a multipurpose cleaning agent. It can be used for washing hands before applying hand sanitizer.

Processes of Liquid Soap Production

When starting the process of making liquid soap, keep in mind the procedures that follow and safety precautions.

1. Process of fermentation

To lessen a chemical's harsh effects, it is fermented by soaking it in water. For an hour or longer.

Caustic soda is an example of a chemical that must be fermented before being utilized to make liquid soap. It will cause irritation if it comes into contact with your skin directly.

2. Process of dissolving

To decrease the size of a chemical's particles, it must be dissolved in water. Some compounds are smaller in size and require dissolution prior to synthesis. It lasts for at least an hour.

3. Blending

Chemicals must be combined, and the mixture must be stirred.

Note that chemicals are mixed into water. It is wrong add water directly to a chemical.

Materials used in the production of liquid soap.

The following are essential for making liquid soap. As follows:

1. **Five Pails**: Useful for fermenting and dissolving important substances.

2. **Stirring stick**: a tool for blending.

3. **A nose mask**: This is worn to avoid breathing in potent substances.

4. Safety gloves are utilized as chemical protection.

5. Measurement cups that have been calibrated are used for measurement.

Chemicals Used In the Manufacture of Liquid Soap

The following substances are required for the production of liquid; it is expected that you avoid exposing your body to them.

1. Nitrosol
In order to thicken liquid soap, the chemical nitrosol is utilized.

2. Caustic soda
In order to erase stains, caustic soda is a chemical used in the manufacture of liquid soap.

3. Soda Ash
Caustic soda is aided by the chemical soda ash, which is utilized in the manufacturing of liquid soap.

4. Texapom

As a foaming agent, texapom is a chemical used in the manufacture of liquid soap. It helps to make foams.

5. Sulfuric Acid

This also serves as a foaming agent.

6. Sodium lauryl sulfate

"SLS" is another name for this. Foam bubbles are made possible by this.

7. Foam bluff

It is a substance that gives soap its claimed capacity to foam.

8. E. D. T. A

This substance (ethylene-diamine-tetra-acetic acid) serves as a preservative. The soap lasts for around six months thanks to it.

9. Water-based color
It gives the combination color.

10. Fragrance
The purpose of fragrance is to provide the mixture odor. The most popular scents to use are apple, lemon, or ambiphur.

11. Glycerin
This serves as a moisturizer in the manufacture of liquid soap.

Manufacturing Liquid Soap

Before beginning to produce liquid soap, keep in mind the following information.

1. Avoid adding water straight to a chemical while it is fermenting or dissolving. To avoid chemical splashes on your skin, it is recommended that you pour water into the pail prior adding chemical.

2. Stir in a single direction while fermenting and dissolving. Don't stir in the opposite direction to prevent chemical splashes on your skin.

3. Don't let kids around chemicals.

4. If you come into contact with a chemical and your skin, wash it right away.

5. Write the names of the chemical mixes in each bucket.

Chemical Requirements and Measurement for a Typical Production of 10 liters

Take the list below to the chemical store and make your purchases if you want to make a regular batch of 10 liters of liquid soap.

- 1/8 kilogram of Nitrosol.
- 1/8 kilogram of Caustic soda.

- 1/2 kilogram of Soda ash.
- 1/8 kilogram Texapom.
- 1 liter of Sulphonic acid.
- 1/8 kg sodium lauryl sulfate (SLS)
- 1 wrap EDTA
- A half-liter foam boaster
- Watery color
- Glycerin

How to Make 10 Liters of Liquid Soap for Handwashing

You can choose to produce your liquid soap for handwashing and distribute to family and friends. It is also beneficial to sell.

To determine the precise amount needed to make 20 liters, multiply the quantity of each component by 2.

1. Label each of the five (5) pails with A through E.
   - 10 liters of water should be poured into the A pail.
   - 1 liter of water should be removed from pail A and placed in pail B.

2. Fill pail B with 1/8 kg of caustic soda.
   - Stir well, then let ferment for at least one hour.

3. Transfer two liters of water from bucket A to bucket C.

- In pail C, add 1/2 kg of soda ash and well stir.
- Leave to ferment for at least one hour.

4. Pour 1 liter of water into pail D after removing it from pail A.

- Fill pail D with 1/8 kilogram of SLS.
- Stir well, then let the mixture dissolve for at least an hour.

5. Transfer a half-liter of water from pail A to pail E.

- To pail E, add 1 wrap of water-based color
- Adequately stir in one direction.

6. Add one E.D.T.A wrap to pail A water.

- Fill pail A with 1/8 kg of nitrosol, and well stir.

- Stir add 1/8 kg of texapom to the pail A mixture.
- Pour 1 liter of sulfuric acid into pail A and thoroughly stir.

7. Add the caustic soda from pail B to pail A and thoroughly stir.

Wait for 30 minutes before packaging for use.